M000198965

ELLIE NEEDS TO GO

OTHER BOOKS ABOUT ELLIE AND TOM

What's Happening to Ellie?
A book about puberty for girls and young
women with autism and related conditions
ISBN 978 1 84905 526 0
eISBN 978 0 85700 937 1

Things Ellie Likes
A book about sexuality and masturbation
for girls and young women with
autism and related conditions
ISBN 978 1 84905 525 3
eIBSN 978 0 85700 936 4

What's Happening to Tom?
A book about puberty for boys and young
men with autism and related conditions
ISBN 978 1 84905 523 9
eISBN 978 0 85700 934 0

Things Tom Likes
A book about sexuality and masturbation for boys
and young men with autism and related conditions
ISBN 978 1 84905 522 2
eISBN 978 0 85700 933 3

Tom Needs to Go
A book about how to use public toilets
safely for boys and young men with
autism and related conditions
ISBN 978 1 84905 521 5
eISBN 978 0 85700 935 7

BY THE SAME AUTHOR

Sexuality and Severe Autism
A Practical Guide for Parents,
Caregivers and Health Educators
ISBN 978 1 84905 327 3
eISBN 978 0 85700 666 0

ELLIE NEEDS TO GO

A book about how to use public toilets safely for girls and young women with autism and related conditions

KATE E. REYNOLDS

Illustrated by Jonathon Powell

Jessica Kingsley *Publishers*
London and Philadelphia

First published in 2015
by Jessica Kingsley Publishers
73 Collier Street
London N1 9BE, UK
and
400 Market Street, Suite 400
Philadelphia, PA 19106, USA

www.jkp.com

Copyright © Kate E. Reynolds 2015
Illustration copyright © Jonathan Powell 2015

All rights reserved. No part of this publication may be reproduced in any material form (including photocopying or storing it in any medium by electronic means and whether or not transiently or incidentally to some other use of this publication) without the written permission of the copyright owner except in accordance with the provisions of the Copyright, Designs and Patents Act 1988 or under the terms of a licence issued by the Copyright Licensing Agency Ltd, Saffron House, 6–10 Kirby Street, London EC1N 8TS. Applications for the copyright owner's written permission to reproduce any part of this publication should be addressed to the publisher.

Warning: The doing of an unauthorised act in relation to a copyright work may result in both a civil claim for damages and criminal prosecution.

Library of Congress Cataloging in Publication Data
A CIP catalog record for this book is available from the Library of Congress

British Library Cataloguing in Publication Data
A CIP catalogue record for this book is available from the British Library

ISBN 978 1 84905 524 6
eIBSN 978 0 85700 938 8

Printed and bound in China

Many thanks to Leigh Robinson, who cared for years for my beautiful long locks, and Heather, Nicky and Bianca, who cared for me when I lost them!

Thanks also to my siblings Shirley, David and Karen, Graeme and Lesley for their support.

Kate

For my grandmother, Daphne Powell, for all her love and encouragement.

Jonathon

A NOTE FOR PARENTS AND CAREGIVERS

Ellie Needs to Go is a book about safety in public toilets for girls and young women with autism or related conditions.

As parents, caregivers or health educators we may find ourselves supporting a child of the opposite gender in a situation where that child needs the toilet in a public place. Teaching 'toilet etiquette' is not something we expect to have to do. Girls typically learn this information by observing others, such as their peers or mothers, but this avenue may not be open to children and young people with autism and related conditions because they may not socialise with non-spectrum others. In addition, mothers may feel uncomfortable with the close and repeated scrutiny autistic children may need to fully understand appropriate behaviours. Typically developing children also are more able to ask specific questions if they are uncertain.

Equipping girls and young women with information about expectations of proper behaviour in public toilets is one plank in ensuring their safety against possible child abuse. If they have knowledge of appropriate behaviour in toilets, they can accurately report any inappropriate behaviours, including sexual advances.

Ellie is in the park with her father.

Ellie needs to use the lavatory.

Ellie has to wait her turn for an empty lavatory.

Women in the line talk to each other.

Ellie enters the first empty lavatory when it is her turn.

Inside the cubicle, Ellie pulls down her underpants and sits on the lavatory.

Ellie is having a period, so she pulls the dirty sanitary pad off her underpants, puts it in a sanitary bag and puts the bag in the sanitary bin.

Ellie has a new sanitary pad in her handbag.
She sticks the new pad in her underpants.

Sometimes Ellie needs to poo.

Ellie also pees when she has a poo, but she knows she should still follow the 'F' rule. That means she wipes her Front First.

When Ellie finishes pooing, she wipes her bottom hole with toilet paper.

Whether Ellie has a pee, a poo or changes her sanitary pad, she pulls up her underpants and makes sure her clothes are on properly before unlocking the lavatory door.

Ellie always washes her hands after she goes to the toilet.

Ellie joins her dad in the park. She's going to have an ice cream!

Ellie joins her dad in the park. She's going to have an ice cream!

ABOUT THE SERIES

Sexuality and sexual safety are often difficult subjects for parents, caregivers and health educators to broach with young people who have severe forms of autism and related conditions. These young people are widely perceived as being 'vulnerable', but the lack of sex education and social opportunities available only increases that vulnerability, leaving them open to child sex and other abuse. Unlike typically developing children who learn by 'osmosis' from their peers, our young people need clear and detailed information provided by those who support them.

This is one of a series of six books – three for girls and young women and three addressing issues for boys and young men. Each book tells a story about the key characters, Ellie and Tom, giving those supporting young women and men something tangible as a basis for further questions from young people. The wording is unambiguous and avoids euphemisms that may confuse readers and listeners. Many young people with severe forms of autism and related conditions are highly visual, so the illustrations are explicit and convey the entire story.

These books are designed to be read with a young person with autism, alongside other more generic reading material.